AIChE Equipment Testing Procedure

Centrifugal Compressors

AIChE Equipment Testing Procedure

Centrifugal Compressors

A Guide to Performance Evaluation and Site Testing

Equipment Testing Procedures Committee
American Institute of Chemical Engineers

AIChE

WILEY

Cover and book design: Lois Anne DeLong

It is sincerely hoped that the information presented in this document will lead to an even more impressive performance by the chemical processing and related industries. However, the American Institute of Chemical Engineers, its employees and consultants, its officers and directors, Equipment Testing Procedures Committee members, their employers, and their employers' officers and directors disclaim making or giving any warranties or representations, express or implied, including with respect to fitness, intended purpose, use or merchantability and/or correctness or accuracy of the content of the information presented in this document. Company affiliations are shown for information only and do not imply approval of the procedure by the companies listed. As between (1) the American Institute of Chemical Engineers, its employees and consultants, its officers and directors, Equipment Testing Procedures Committee members, their employers, and their employers' officers and directors, and (2) the user of this document, the user accepts any legal liability or responsibility whatsoever for the consequences of its use or misuse.

A Joint Publication of the Center for Chemical Process Safety of the American Institute of Chemical Engineers and John Wiley & Sons, Inc.

Published by John Wiley & Sons, Inc., Hoboken, New Jersey

Published simultaneously in Canada

Library of Congress Cataloging-in-Publication Data:

American Institute of Chemical Engineers
AIChE Equipment Testing Procedure : Centrifugal compressors / a guide to performance evaluation and site testing analysis.
p. cm.
Includes index.
ISBN 978-1-118-62781-5 (paper)

10 9 8 7 6 5 4 3 2 1

AMERICAN INSTITUTE OF CHEMICAL ENGINEERS EQUIPMENT TESTING PROCEDURES COMMITTEE

Chair: **Rebecca Starkweather, P.E.**
Scientex, L.C.

Vice-Chair: **Prashant D. Agrawal, P.E.**
Consultant

Past Chair: **James Fisher, P.E.**
Amgen, Inc.

CENTRIFUGAL COMPRESSOR PROCEDURE SUBCOMMITTEE

Chair: **Royce N. Brown**
RNB Engineering

Vice-Chair: **Mark Cooper**
Lauren Engineers & Constructors, Inc.

General Committee Liaison: **S. Dennis Fegan**
Powerdyne Consultants

Working Committee

Mary Ann Strite
Elliot Group (Retired)

Gary M. Colby
Dresser-Rand Inc.

Company affiliations are shown for information only and do not imply Procedure approval by the company listed.

CTOC Liaisons

Bob Hoch
Retired

Shrikant Dhodapkar, PhD
Dow Chemical

AIChE Staff Liaison: **Stephen Smith**

General Committee Members at Publication

David Dickey, PhD
Mixtech, Inc.

S. Dennis Fegan
Powerdyne, Inc.

P.C. Gopalratnam, PhD, P.E.
INVISTA

Robert McHarg, P.E.
UOP (Retired)

Richard O'Connor, P.E.
R.P. O'Connor & Associate

Neil Yeoman, P.E.
Retired

General Committee Past Members

Thomas Yohe, P.E.
Yohe Consulting

Rajinder Singh
Purestream Technology

John Kunesh, PhD, P.E.
Deceased

Dick Cronin, P.E.
E. I. duPont

Joe Hasbrouck, P.E.
R.P. O'Connor & Associates

Tony Pezone, P.E.
E. I. duPont (Retired)

Table of Contents

List of Tables

List of Figures

100.0 PURPOSE AND SCOPE

101.0 Purpose

This testing procedure suggests methods of conducting and interpreting field tests on centrifugal compressors. Measuring methods practical for field test conditions are presented to guide the user. These tests may be desired for the following purposes:

- To verify the guaranteed performance operating conditions;
- To determine how to improve performance when the compressor does not perform as required or does not satisfy process requirements;
- To ascertain if performance problems are being induced by the installation;
- To track the condition of the compressor internals, especially if the gas is corrosive or causes fouling, or if a process upset has occurred that might damage the compressor;
- To provide guidelines for the user in modifying existing equipment; and,
- To enable the user to estimate the gas composition through comparisons to the shop test results.

101.1 If the data is solely for the user, adherence to code recommendations is up to him/her, and a flow and a discharge pressure reading may be sufficient.

101.2 When the manufacturer is brought into the picture, more stringent adherence to good testing practices is required, especially for acceptance testing or for correcting performance deviations.

101.3 This document is not necessarily intended to be a purchasing specification.

102.0 Scope

102.1 The scope of the procedure is limited to performance (aerodynamic) testing of centrifugal compressors at the job site. It does not address performance testing of fans, blowers, reciprocating, axial, screw or other types of compressors.

102.2 It should be noted that the preferred method of establishing centrifugal compressor performance is under controlled laboratory conditions. Testing can generally be accomplished by the compressor manufacturer at the factory and may be witnessed by the purchaser. It should also be noted that this field test procedure should not be used when test code accuracy is required.

102.3 Mechanical testing, such as monitoring vibration and bearing temperatures, is not included in this specification.

102.4 Material and other component tests that are generally associated with quality assurance are also excluded.

102.5 Limitations

102.5.1 The procedures are not designed to serve as a legal code for "acceptance tests." Procedures provide instructions for measuring performance without regard to expected design or calculated performance.

102.5.2 Wherever possible, procedures use approved AIChE nomenclature and symbols. If special nomenclature or symbols are necessary, all deviations will be reviewed and approved by the appropriate AIChE organization.

102.5.3 Procedures will not typically include any detailed design methods. If this information is necessary for proper Procedure understanding, it will be included in the Appendix.

200.0 DEFINITIONS AND DESCRIPTION OF TERMS

This section contains definitions for all terms used in this procedure. Nomenclature and typical units are also listed in the Appendix (Section 800.0, pgs. 40-41) The name of the term is given in italics, followed by the symbol in parenthesis, and then the definition. Note that many of the symbols are case sensitive. Where applicable, equations have been used in the definitions.

Terms used for calculations should be defined so they are readily understood. It is better to err by using too many subscripts than not enough. If subscripts cannot readily be added, such as with some computer software, the subscript can be changed to regular text (e.g. t_1 can be written as t1).

ASME PTC-10 uses *Work* where most other references, such as API, use *Head*; in that reference, *Work* (e.g. Head/η or h_2 - h_1) is called *Work Input*. The more familiar *Head* and the better defined *Work Input* are used here. (Note that in some equations in the book, the symbol W_p will be used instead of H_p.)

200.1 Absolute Pressure (p)

Absolute Pressure is measured from absolute zero pressure, or from an absolute vacuum. It equals the sum of the atmospheric pressure and gauge pressure. Units of measure are psia. Pressures that are not identified as gauge pressures or labeled psig in this paper are absolute pressures.

$$P = P_{gauge} + P_{atm} \qquad \textbf{(200.1.1)}$$

200.2 Absolute Temperature (T)

Absolute Temperature is the temperature above absolute zero. It is equal to the degrees Fahrenheit plus 459.7 and is stated as degrees Rankine.

$$T\ °R = t\ °F + 459.7 \qquad \textbf{(200.2.1)}$$

200.3 Acoustic Velocity (a)

Acoustic Velocity is the speed of sound in a gas at a given pressure and temperature.

$$a = \sqrt{k\ g\ R\ T\ z} \qquad \textbf{(200.3.1)}$$

200.4 Area (A)

Area of a pipe or other item, in square ft (ft_2)

200.5 Bearing Loss ($P_{bearings}$)

This term describes the mechanical power loss in a bearing, which is exhibited as heat dissipation, noise and vibration. Bearing loss is usually totalized for a compressor case or section (i.e., journal and thrust bearing losses combined).

200.6 Bypass Control

Bypass control is a control system whereby some or all of the gas flow through a compressor stage or section is routed back to the inlet of a section or stage of the compressor through a control valve and gas cooler. The purpose of this system is to provide turndown capacity or total recycle during periods of startup, shutdown or reduced compressor capacity.

200.7 Capacity (q)

This term represents the volume rate of flow of gas compressed and delivered dependent on the conditions of stagnation pressure, stagnation temperature, and gas composition prevailing at the compressor inlet. May also be shown as *Q*. Capacity is frequently call *Volume Flow Rate* to differentiate it from *Mass Flow Rate (w)*.

200.7.1 *Inlet Cubic Feet Per Minute (ICFM) and Actual Cubic Feet Per Minute (ACFM)*

These terms refer to inlet capacity, in cubic feet per minute, determined at suction (compressor inlet flange) conditions of pressure, temperature, compressibility, moisture and gas composition. ACFM and ICFM are abbreviations for actual and inlet cubic feet per minute, which are identical.

200.7.2 *Standard Cubic Feet Per Minute (SCFM)*

SCFM is an abbreviation for capacity in standard cubic feet per minute of gas at 14.7 psia and 60 ° F, with a compressibility factor of 1.0. Since the definition of SCFM varies between users and industries, it is recommended that users check the definition and state the specific defining terms being used.

200.7.3 *Choke Flow*

Also called *Stonewall*, this is a measurement of the maximum capacity that a stage or a compressor section can handle at specified speed and flow conditions. The head curve will drop off sharply and become vertical. Choke usually occurs at high flows for stages operating at high Mach numbers. It can occur under different conditions, as with a poor impeller design, selection, or manufacture. It should be noted that the limits of centrifugal compressor capacity may be caused by inefficiencies that occur as the aerodynamic components become mismatched at high flows or by losses. Although the head curve will not drop off as sharply, it can be hard to distinguish between choke flow and a loss curve.

200.8 Compressibility Factor (z)

Compressibility Factor relates the true density or specific volume of a gas to the corresponding values calculated assuming ideal gas behavior. Technically, z is 1.0 for an ideal gas.

200.9 Critical Pressure (P_c)

This measure is the highest pressure under which distinguishable liquid and vapor phases can exist in equilibrium. The operating pressure divided by the critical pressure is equal to the Reduced Pressure.

$$P_r = \frac{P}{P_c} \qquad (200.9.1)$$

The reduced pressure and reduced temperature are used to find the compressibility factor z on compressibility charts.

200.10 Critical Temperature (T_c)

Critical Temperature is the highest temperature under which distinguishable liquid and vapor phases can exist in equilibrium. See the discussion on **Critical Pressure**.

$$T_r = \frac{T}{T_c} \qquad (200.10.1)$$

200.11 Density (ρ)

Density is the mass of fluid per unit volume and is the inverse of the *Specific Volume.*

$$\rho = \frac{1}{v} \qquad (200.11.1)$$

200.12 Diameter (D)

The upper case *diameter* (D) term will, in general, refer to an impeller outer diameter or a pipe diameter. The lower case diameter (d) is the throat diameter of a flow measuring device.

200.13 Diaphragm

This term describes a compressor component that fits into the multistage compressor casing and forms the stationary flow paths of a stage.

200.14 Diffuser

The *diffuser* refers to the portion of the stage from the impeller tip to the return channel in which velocity is converted to pressure. It may or may not have vanes.

200.15 Dimensionless Curves

This term is used to describe compressor performance curves of the head coefficient μ, work input coefficient γ, and efficiency η, as a function of the flow coefficient ϕ; a semi- dimensionless term q/N may be used instead of the flow coefficient.

200.16 Driver

The *driver* is a power delivery device that connects to, and turns the shaft of, a compressor. The driver may be a steam or gas turbine, a gas expander, or a motor.

200.17 Dry Bulb Temperature (t_{DB})

This term defines the temperature of a gas flowing around a thermocouple or other temperature sensing device measured with no water evaporation occurring at the sensing point. Unless stated otherwise, temperatures used in this procedure are dry bulb temperature.

200.18 Efficiency (η)

Efficiency describes the ratio head to the work input (supplied by a driver). Efficiency may be adiabatic or polytropic, depending on whether an adiabatic or polytropic head is used. The ratio is used for calculations, and may be multiplied by 100 to express the efficiency as a percentage value.

$$\eta = \frac{Head}{Work\ Input} \qquad (200.18.1)$$

200.19 Enthalpy (h)

Enthalpy is the internal energy and pressure-volume energy of a gas at a given pressure and temperature.

200.20 Entropy (s)

For a closed thermodynamic system, *Entropy* is a quantitative measure of the amount of thermal energy not available to do work.

200.21 Fan Laws

This term describes the relationships between speed, capacity, power requirement and the developed head for a fan or rotating element of a compressor. Fan laws may be used with an adiabatic or polytrophic head.

$$q \cong N$$

$$HEAD \cong N^2$$

$$HP \cong N^3 \qquad (200.21.1)$$

200.22 Flow Coefficient (ϕ)

Flow Coefficient is a dimensionless factor that relates the suction capacity of an impeller to its operating speed and diameter.

$$\phi = \frac{700q}{N d_2^3} \qquad (200.22.1)$$

200.22.1 Flow coefficient is a design term, and values run from 0.01 (and smaller) for low flow impellers, to greater than 0.20 for very high flow impellers.

200.22.2 If the data for several impellers is plotted as flow coefficient, head coefficient, and work coefficient on log-log paper, the curves will be similar to each other. The similarities of the stages can be observed by overlaying the curves.

200.23 Gas Constant (*R*)

This describes the constant used in the gas law.

$$pv = zRT$$

$$R = \frac{R_u}{M} \qquad (200.23.1)$$

200.24 Gauge Pressure (*p*)

Gauge pressure is the pressure above or below atmospheric pressure, and the units are psig (See Absolute Pressure). Any gauges that read Absolute Pressure will be so labeled. Pressures that are not identified as gauge pressures or labeled psig in this paper are absolute pressure.

200.25 Gear Loss (P_{gear})

The mechanical power loss in a speed increaser or reducer, which is exhibited as heat dissipation, noise, and/or vibration is defined as *Gear Loss*.

200.26 Gravitational Acceleration (*g*)

This term describes the free fall acceleration experienced by a falling body near the earth's surface. By International Convention, the standard value of gravitational acceleration is 32.174 ft. per sec per sec at sea level and 45 degrees North latitude (9.8066 meters per sec per sec).

200.27 Gravitational Dimensional Constant (g_c)

This dimensional constant is derived from the system of units used wherein mass is expressed as pounds mass (lb), and force is expressed as pounds-force (lbf). Its value is 32.174 lb_m -ft per $lb_f\ sec^2$ and is unaffected by local gravitational force.

200.28 Horizontal Split Case

This phrase describes a type of compressor design where the casing is horizontally split at the centerline.

200.29 Impeller

An *impeller* is the rotating element attached to the compressor shaft that imparts work to the gas in the form of velocity, which is then converted to pressure.

$$U = \frac{\pi D N}{720} \qquad (200.29.1)$$

200.30 Impeller Tip Speed (*U*)

This is the measure of the tangential velocity at the outer diameter of an impeller.

200.31 Inlet Guide Vanes

Vanes located at the inlet to an impeller. Three types are used:

200.31.1 *Gas Stream Line Straightening* vanes

200.31.2 *Adjustable or Variable Inlet Guide* vanes, on which the angle of the vanes can be varied by a controller to cause prewhirl or antiwhirl, decreasing or increasing the head and work input of the stage.

200.31.3 *Fixed Prewhirl or Antiwhirl Vanes.* These are installed as part of the return channel of multistage compressors. This type of vane was popular prior to the 1960s.

200.32 Isentropic Compression

This term describes a process in which the entropy is constant and follows a constant entropy line on a Mollier chart. For compressors, it refers to reversible *Adiabatic Compression.*

200.33 Isentropic Head (H_s)

This term refers to the measure of the head required to compress a unit mass of gas in an Isentropic compression process from the inlet pressure and temperature to the discharge pressure. It should be noted that ASME PTC-10 uses *Work* for *Head* and *Work Input* for *Work.*

200.34 Isentropic Head Coefficient (α_s)

This coefficient is the product of the Isentropic head developed by a stage, and the gravity constant, divided by the square of the impeller tip speed. For a section, use the total Isentropic Head of all impellers and the sum of the square of tip speed of each impeller.

200.35 Isentropic Volume Exponent (n_s)

This measure is the logarithm of the absolute outlet to inlet pressure ratio divided by the logarithm of the inlet to outlet volume ratio for an isentropic compression path passing from inlet to outlet conditions

$$n_s = \frac{\ln(p_d / p_i)}{\ln(v_i / v_d')}$$

where υ_d is the specific volume at the discharge pressure and the inlet entropy.

200.36 Machine Mach Number (M_m)

This term is defined as the ratio of the velocity at the outer diameter of the first impeller to the acoustic velocity in the gas at the inlet conditions of stagnation pressure and temperature. The Mach number used in impeller design and selection is based on the area at the inlet of the impeller, rather than the discharge. The Machine Mach number is an artificial value and, consequently, may have a value greater than 1.

$$M_m = \frac{U}{a_i} \qquad \textbf{(200.36.1)}$$

Where a_I is the acoustic velocity at the impeller inlet conditions.

200.37 Machine Reynolds Number (*Re*)

For this process, *Re*, or the *Machine Reynolds Number* is defined by the equation where *U* is the velocity at the outer diameter of the first impeller, υ is the kinematic viscosity of the gas at the compressor inlet stagnation conditions, and *D* is the exit width of the first stage impeller. These variables must be expressed in consistent units so as to yield a dimensionless ratio.

$$\mathrm{Re} = \frac{UD}{\upsilon} \qquad \textbf{(200.37.1)}$$

200.38 Mass Flow Rate (*w*)

Mass Flow Rate is the amount of mass throughput per time element in lb_m/min. Compressor performance guarantees are presented in terms of mass flow rate rather than volume flow rate *q*. Volume flow rate is subject to change, due to the use of inlet throttling or changes in the inlet pressure on multi section compressors. Compressor performance, however, is a function of volume flow.

$$q = wv = \frac{w}{\rho} \quad \textbf{(200.38.1)}$$

200.39 Maximum Allowable Temperature

This measurement is the maximum continuous temperature for which the manufacturer has designed the equipment (or any part to which the term is applied) when handling the specified fluid at the specified pressure.

200.40 Maximum Continuous Speed

Measured in revolutions per minute (rpm), *Maximum Continuous Speed* is the speed at least equal to the product of 105%, or the highest speed required by any of the specified operating conditions.

200.41 Molecular Weight (*M*)

This term describes the weight of a molecule of a substance in reference to that of an atom of carbon at 12.000.

200.42 Mollier Diagram

The *Mollier Diagram* is an illustration that depicts gas properties.

200.43 Normal Operating Point

This term specifies the operating point at which usual operation is expected and optimum efficiency is desired.

200.44 Normal Speed

The speed corresponding to the requirements of the normal operating point is designated *Normal Speed.*

200.45 Nozzle

This term defines the flanged pipes on the compressor casing that pass gas flow into or out of a compressor section or stage.

200.46 One Hundred (100) Percent Speed

Should higher than normal speeds be required to satisfy other specified operating points, the highest of such speeds is designated the *100 percent speed.* The compressor rated point is on the 100 percent speed curve at a capacity equivalent to the highest specified capacity point. If there are no other such specified operating points, the 100 percent speed will be treated as the normal speed. The 100 percent speed of motor driven compressors will be equal to the gear ratio (if any) times the full load speed of the motor furnished.

200.47 Pipe Reynolds Number (RD)

For gas flow in a pipe, as would be the situation in the flow metering calculations, the *Reynolds number* is defined as:

$$R_D = \frac{VD}{\upsilon} \qquad (200.47.1)$$

where the velocity V, characteristic length D, and kinematic viscosity υ, are to be used as follows: velocity is the average velocity at the pressure measuring station, the characteristic length is the inside pipe diameter at the pressure measuring station, and the kinematic viscosity is that which exists for the static temperature at the temperature measuring station. The variables in the Reynolds number must be expressed in consistent units so as to yield a dimensionless ratio.

200.48 Polytropic Compression

This term describes a reversible compression process between the compressor inlet and discharge conditions that follow a path such that, between any two points on the path, the ratio of the reversible work input to the enthalpy rise is constant.

200.49 Polytropic Exponent (*n*)

This value is derived when the logarithm of the absolute outlet to inlet pressure ratio is divided by the logarithm of the inlet to outlet volume ratio for any irreversible gas compression path passing from the inlet to outlet conditions.

$$n = \frac{\ln r_p}{\ln r_v} = \frac{\log_{10} r_p}{\log_{10} r_v} \qquad (200.49.1)$$

The following expression is true only for an ideal gas (where $z = 1.0$):

$$n = \frac{\ln r_p}{\ln r_v} \qquad (200.49.2)$$

The following expression is true only if the specific heat at constant volume $c\text{-}v$ is nearly constant:

$$\frac{n}{n-1} = \eta_p\left(\frac{k}{k-1}\right)$$

$$n = \frac{1}{Y - m(1+X)}$$

$$(200.49.3)$$

The most rigorous definition of n is given by ASME PTC-10:

$$n = \frac{zR}{I}\left(\frac{1}{p} + X\right) \qquad (200.49.4)$$

where X and Y are compressibility functions called *Schultz Functions.*

200.50 Polytropic Head (H_P)

This term represents the reversible head required to compress a unit mass of the gas in a polytropic compression process. Several equations are used to define polytropic head, but, the most frequently used configuration is as follows:

$$H_p = RT_i z\left(\frac{n}{n-1}\right)\left(\left(\frac{P_d}{P_i}\right)^{\frac{n-1}{n}} - 1\right) \qquad (200.50.1)$$

To use this equation correctly, care must be used in determining the *Polytropic Exponent n* defined above. This equation is a handy tool for quickly approximating the polytropic head.

$$H_p = 144\left(\frac{n}{n-1}\right)p_i v_i (p_d v_d - p_i v_i) \qquad (200.50.2)$$

Polytropic head can also be expressed as:

$$H_p = 144\left(\frac{n}{n-1}\right) f\, p v_i \left(\left(\frac{P_2}{P_1}\right)^{\frac{n-1}{n}} - 1\right) \qquad (200.50.3)$$

The most rigorous equation for polytropic head is used by ASME PTC-10. Please note that the symbol W_p can also be used to represent this term.

200.51 Polytropic Head Coefficient (μ_p)

This term describes a dimensionless factor that, when multiplied by the square of the tip speed and divided by the gravitational constant, equals the polytropic head developed for a specific impeller.

$$\mu_p = \frac{H_p}{\sum\left(\frac{U^2}{g}\right)} \qquad (200.51.1)$$

200.52 Polytropic Head Factor (*f*)

This term describes another dimensionless factor that can be defined as:

$$f = \frac{h_d' - h_i}{\left(\frac{n_s}{n_s - 1}\right)\left(p_d v_d' - p_i v_i\right)144} J \qquad (200.52.1)$$

The terms h_d and v_d are the enthalpy and specific volume at the discharge pressure and the inlet entropy, and n_s is the Isentropic Volume Exponent.

For most gases and operating conditions, f is approximately unity; if the value varies more that a percent or two from unity, gas properties should be checked carefully.

200.53 Power (*P*)

Power is the work required to drive the compressor. Compressor gas power is computed as:

$$P_g = \frac{Hw}{3300\eta} \qquad (200.53.1)$$

200.53.2 The compressor power is the gas power plus the mechanical losses, which includes the bearing and seal losses.

$$P_{comp} = P_g + P_{bearings} + P_{seals} \qquad (200.53.2.1)$$

200.53.3 The power supplied by the driver includes the power required by all compressors and gears in the train.

$$P_{train} = \sum P_{comp} + \sum P_{gear} \qquad (200.53.3.1)$$

200.54 Pressure Ratio (r_p)

Pressure ratio is the ratio of the (absolute) discharge pressure to the (absolute) inlet pressure of a section of a compressor. For a compressor, r_p is usually greater than unity.

$$r_p = \frac{p_d}{p_i} \qquad (200.54.1)$$

200.55 Schultz Functions (*X, Y*)

These are compressibility functions used in the calculation of the *Polytropic Exponent* ***n***.

200.56 Seal Losses (P_{seals})

This term describes the mechanical power loss in the shaft end seals, which is exhibited as heat dissipation, noise, and vibration. *Seal loss* is usually totalized for a compressor case or section; the power loss occurs primarily in oil-fed seals. Other seals, such as carbon ring, labyrinth, and dry gas seals, have such low losses that this factor is not counted. The losses in interstage seals result from flow that is recirculated, either around each stage through the interstage seals or from the compressor discharge through the balance piston seal to another part of the compressor. These losses are inherent in the compressor performance and should not be added in separately.

200.57 Sidestream (Sideload)

Sidestream or *Sideload* is a gas stream entering or leaving a compressor between two sections. Another term used is *Extraction*, but it denotes only exiting gas streams.

200.58 Specific Gravity (G)

When measuring gases, *Specific Gravity* is the ratio of the density of the gas at a pressure of 14.7 psia and a temperature of 68 °F to the density of dry air at the same pressure and temperature and an apparent molecular weight of 28.970. For liquids, it is the ratio of the density of the liquid at 60 °F to the density of pure water at 60 °F.

200.59 Specific Heat at Constant Pressure (c_p)

This term represents the change in energy of a gas per unit or mass held at constant pressure for a change in temperature.

200.60 Specific Heat at Constant Volume (c_v)

This term describes the change in energy of a gas per unit of mass held at constant volume for a change in temperature.

200.61 Specific Heat Ratio (k)

Specific Heat Ratio is the ratio of the specific heat at constant pressure to the specific heat at constant volume for any gas.

$$k = \frac{c_p}{c_v} \qquad (200.61.1)$$

200.62 Specific Volume (v)

This measure is the volume occupied by a unit mass of the fluid.

$$v = \frac{RTz}{144p} \qquad (200.62.1)$$

where T, z and p are at the same location, such as the inlet.

200.63 Speed (N)

This describes the rotative speed.

200.64 Stability

Stability is the percent change between the rated capacity and the surge point at rated speed.

200.65 Stage

Stage is a portion of a compressor containing an impeller, diffuser, and return channel or volute.

200.66 Stagnation (Total) Pressure (p or p_t)

Compressor performance is based on *Stagnation* or *Total Pressure*. Usually measured by an impact tube, it is the pressure that would be measured at the point when a moving gas stream is brought to rest and its kinetic energy is converted to an enthalpy rise by an isentropic compression from the flow condition to the stagnation condition. In a stationary body of gas, the static and stagnation pressures are numerically equal. The *Stagnation Pressure* is equal to the *Static Pressure* plus the *Velocity Pressure*.

$$p_t = p_s + p_v \qquad \textbf{(200.66.1)}$$

Unless otherwise stated, a pressure noted as >*p*=—without a subscript—is the total or stagnation pressure; the subscript >*t*= is used where there may be confusion between *Total (Stagnation)* and *Static Pressures*. Static pressures are, in general, only used with test and design calculations.

200.67 Stagnation (Total) Temperature (T)

This defines temperature that would be measured at the stagnation point if a gas stream were brought to rest and its kinetic energy converted to an enthalpy rise by an isentropic compression process from the flow condition to the stagnation condition. It is difficult to measure a true total temperature; the difference between static and total temperatures is small and may be calculated empirically or ignored.

200.68 Static Pressure (P_s)

This term is the pressure in the gas measured such that no effect is produced by the velocity of the gas stream. It is the pressure that would be shown by an instrument moving at the same velocity as the moving stream, and is used as a property in defining the thermodynamic state of the fluid.

200.69 Static Temperature (T_s)

Similar to Static Pressure, *Static Temperature* is what would be shown by a measuring instrument moving at the same velocity as the fluid stream.

200.70 Temperature Ratio (R_t)

This measure is the ratio of the absolute (in °R) discharge temperature to the absolute inlet temperature for a compressor section. For a compressor, R_t is usually greater than unity.

$$r_t = \frac{T_d}{T_i} \quad (200.70.1)$$

200.71 Turndown

This describes the percent change between the rated capacity and the surge point at rated head when operating at design suction temperature and gas composition.

200.72 Universal Gas Constant (R_u)

This term is a physical constant for gases that relates pressure, specific volume and temperature.

200.73 Velocity Pressure (p_v)

The stagnation pressure minus the static pressure in a gas stream equals the *Velocity Pressure*. It is the differential pressure reading of a pitot tube. If the velocity pressure is fairly uniform across the pipe, it can be calculated from the pipe size, the volume flow at that point, and the pressure:

$$V = \frac{q}{60A}$$

$$p_v = \frac{\left(\frac{V^2}{2g}\right)}{144p_s}$$

where A is the area of the pipe or flange where the static pressure is measured.

200.74 Viscosity (υ)

Viscosity is the resistance to deformation offered by any real fluid subjected to a shear stress.

200.75 Volume Flow Rate (q) (see *Capacity*)

200.76 Volume Ratio (r_v)

This is a measure of the ratio of the specific volume of the gas at the inlet stagnation conditions to the specific volume of the gas at the discharge stagnation conditions.

Note that the *Volume Ratio* is expressed as the inlet quantity over the discharge quantity, whereas temperature and pressure ratios are expressed as the discharge quantity over the inlet. For a compressor that is not operating at choke flow, r_v, r, and r_t are all greater than one.

200.77 Wet Bulb Temperature (t_{WB})

This temperature measure is taken during humidification of water into a gas stream.

200.78 Work Input (W)

Work input is the enthalpy rise across the compressor.

200.79 Work Input Coefficient (γ)

This coefficient is a dimensionless factor that, when multiplied by the square of the tip speed divided by the gravitational constant, equals the work input developed for a specific impeller.

$$\gamma = \frac{h_2 - h_1}{\sum\left(\frac{u^2}{g}\right)} \qquad (200.79.1)$$

300.0 TEST PLANNING

301.0 General Guidance

There are three ways to discuss site, or field, aerodynamic (performance) testing. The shortest way is to use a test code, such as ASME PTC-10, and follow it exactly. The next shortest is to list the steps that should be followed, without any explanation of why things should be done as stated. Besides causing frustration, these two methods are likely to result in poor or useless data – and/or unnecessary expense.

The third alternative, and the method used here, is to explain what should be done, suggest what might be eliminated or modified, point out the most likely problems, and discuss the reasoning.

302.0 Communication

Unless one person does all of the testing and data reduction, communication is important. Even if performed by a single person, he or she may appreciate notes on the events that occur and his/her reasoning at the time. Communication is even more important when the personnel involved are separated in time, location–and language.

It is frustrating to struggle for hours or days with a data point that contradicts other results only to learn that, when the data was acquired, it was not thought to be a good point. If the data is read from several sets of instruments (e.g., at the compressor and in the control room), is one set considered more representative than the other?

Events such as the pipes vibrating or emitting different noises at certain points should be noted. If there are duplicate compressors, do you think one performs better than the other, even if you can't prove it?

The other people working on the test may not care when the data collector had his coffee break, but they would like most of the test details.

400.0 INSTRUMENTS AND METHODS OF MEASUREMENT

401.0 Minimum Instrumentation for Checking Compressor Performance

How much instrumentation is required to check compressor performance? The answer varies with the compressor design. Multi-section compressors require more instrumentation than a single section compressor. The instrumentation listed below does not satisfy PTC-10 requirements and, hence, is unlikely to settle performance disputes. It does, however, provide a starting point.

401.1 For a single section compressor (one inlet and one discharge), the following is required:

- Gas analysis: If the gas is air or another pure gas, this is easy.
- Speed of the compressor: The driver speed with the gear ratio is satisfactory.
- Barometer for low pressure compressors ($P1 < 100$ psia) or an altitude for estimating the barometric pressure for higher pressures.
- Inlet pressure
- Inlet temperature
- Discharge pressure
- Mass flow rate
- Discharge temperature

Note that at least one pressure or temperature measurement is required at each location; two or more is better.

401.2 Head can be calculated within a couple percent without a discharge temperature. A discharge temperature or the power consumption is required to determine the efficiency.

402.0 The Ideal Beginning

Ideally, the first steps for a site aerodynamic test of a compressor are taken before the compressor is purchased! Unless the compressor is a "standard" design or an exact duplicate of a tested unit, the compressor should be shop tested.

402.1 The site should be designed for future testing by having adequate straight runs of pipe, thermowells for thermocouples, connections for pressure readings, a flow measuring device, and the means for acquiring gas samples to be analyzed, either on site or elsewhere. Provisions should also be made for the installation of the measurement equipment needed to gather the data cited above in 401.1, and the discharge temperature.

402.2 Many newer field installations are built with some instrumentation already in place. Use of existing instrumentation should be considered in the planning of the performance test.

402.3 Special instrumentation for measuring temperatures at the end of a section should be installed in compressors with sidestreams.

403.0 Schematic of Test Piping and Instrumentation

The initial step in planning a performance test is drawing a schematic of the test piping and the instrumentation. This schematic should show only the items that pertain to the test, not the entire plant. However, in the schematic, it is better to show too much detail than too little. Instrumentation locations should be noted, and all instrumentation identified.

As a rule of thumb, the schematic should show the piping within 20-30 pipe diameters of all compressor inlets and discharges and the flow measuring device(s). Other items that should be shown in the schematic are:

- Valves, elbows, separation drums, silencers, flow straighteners, coolers
- Changes in pipe diameter
- Secondary flows entering the piping near the compressor
- Lines for bypasses or recirculation around the compressor

All of these items cause pressure drops or induce swirl in the flow that can penalize the compressor performance if they are not properly addressed.

403.1 Note the approximate length of the straight runs of piping.

403.2 The schematic should be updated as the planning continues to show changes and additions.

404.0 When Should Tests Be Conducted?

If the testing is being done to verify the guarantee performance, the manufacturer will want the test conducted within the first five minutes of running! However, unless the compressor is subjected to conditions that cause instant performance deterioration, this demand is unreasonable and unnecessary. A more reasonable time is one to two months after installation. If the compressor performance appears deficient, the testing should be repeated after the compressor has been opened, inspected, and certified to be in "nearly new" condition. In this instance, the new testing should be done immediately after start-up.

405.0 Test Planning

The test should be scheduled at a time when suitable operating conditions can be obtained and the presence of key personnel to carry out the test can be assured. Enough time should be allowed for the test team to determine objectives and provide proper instrumentation before the actual test is performed.

405.1 Test Codes

405.1.1 A test code dictates the best way for conducting a test, based on the experience gained from a wide range of tests performed by a number of people at numerous test facilities.

405.1.2 The parties involved in conducting any test need to establish beforehand all the applicable codes under which the tests are being conducted, as well as the specific areas where such code requirements cannot be met.

405.1.3 ASME PTC-10-1965 is recognized in the United States as the primary reference for the performance testing of centrifugal and axial compressors. Since other test codes contain similar material, it is used as the reference for this paper and is called PTC-10. This code concentrates on testing at the manufacturer's facilities, otherwise known as the shop test.

405.1.4 Ideally, all tests would be set-up according to the test code guidelines; however, most shop tests include some judicious "adjustments" of the Code guidelines. Instead of expecting site tests to follow a code to the letter, the manufacturer and the user should work together to determine the compromises that must be made—or those that are simply unavoidable—and their effects on the test results.

405.1.5 Test codes are very rigorous because the penalty for not meeting the guaranteed performance can be quite high, both in money and time. A primary goal of shop testing is to run acceptable tests as quickly and economically as possible, so the compressors can be shipped. Adequate test results can be obtained without meeting every code guideline.

405.1.6 While codes like PTC-10 are the authorities on performance testing, they contain condensed information that is hard to find and interpret. Regular users argue about some items.

405.1.7 Other applicable test codes are listed in Appendix 801.

405.2 Equipment

405.2.1 After determining which measurements are necessary for achieving the test objectives, the appropriate equipment and/or instrumentation must be selected and installed.

405.2.2 The required precision or accuracy of the measurements is an important factor in determining which instruments should be used. Also, proper consideration must be given to calibrating the instruments and installing them according to recognized procedures or practices.

405.3 Process Considerations

405.3.1 If the site performance test is being conducted to verify the compressor meets the guarantee, the operations must be very close to the specified conditions. Table 3 of PTC-10 is often cited as the allowable limits. However, for some compressors, the > allowable limits = are too lax. This is especially true for the inlet temperature.

405.3.2 If the test is to be performed under actual process conditions, as in an operating production unit, attention must be given to the possible restrictions.

405.3.2.1 In many processes, flow cannot be varied at will.

405.3.2.2 If a bypass is used, care must be taken to ensure the bypassed flow does not upset the flow, inlet pressure, and inlet temperature measurements.

405.4 Safety

405.4.1 Since this procedure is directed towards site testing, it is assumed that the compressor will be running on the design gas at or close to the design conditions. That is, the compressor will be operating at the conditions for which it was designed. For site testing on the design gas, plant safety rules are generally more stringent than those at a manufacturer's shop.

405.4.2 Any equipment testing must conform to the latest requirements of all applicable safety standards. These include, but are not limited to, plant, industry, Local, State, and Federal regulations. It is recommended that all testing be conducted under the supervision of personnel fully experienced in plant and equipment operating practices.

405.5 Environmental Considerations
The test procedures must conform to the latest requirements of all applicable environmental conditions regulating equipment in normal operation.

405.6 Pre-test Inspection of Physical Facilities
Prior to the start of testing a compressor, the test engineer should oversee the following activities:

405.6.1 A carefully conducted inspection of the entire test arrangement should take place. Those persons who will make test observations should be part of the inspection team in order to become familiar with the equipment layout and the location of the test instruments to be observed.

405.6.2 All test probes, such as thermocouples and pitot tubes, should be installed and connected to their respective readout instruments.

405.6.3 Pressure gauges, clearly tagged as to in what location they belong, should be installed and connected to appropriate test points. Gauges should be of the appropriate range for the expected test conditions.

405.6.4 Piping valves and fittings should be thoroughly examined and approved as suitable for the expected test pressure and temperature.

- All relief valves must be in place and tagged for the set relief pressure. Piping on valves should lead to safe discharge points.
- Relief valve supports should be adequate to withstand the stresses and vibration that may occur during venting.
- If used, check valves should be installed in the proper direction for normal flow.
- All control valves, either manual or automatic, must be installed in the proper direction.
- Automatic valves should be stroked using their respective controllers to see that they respond with the proper action.

405.6.5 All orifice plates or flow nozzles must be installed in the proper direction in gas meter runs. If required, flow equalizers and straighteners should be installed.

405.6.6 All trip and alarm circuits should be tested and adjusted immediately before compressor startup.

405.6.7 Cooling water systems to exchangers should be piped up and ready to operate.

406.0 Conducting the Test

406.1 Shop test data is acquired in a single run, which may last more than one day, with the flow being varied to give data from surge to stonewall. Flow can be easily varied in the shop test.

406.2 If the system can be varied, site test data should be taken at several flow points. The problem with a single flow point is that the performance curve can have any slope through that point, including vertical, which is choke or stonewall. The single flow point makes it impossible to judge trends or to sort out bad data items.

406.3 If flow is bypassed, the schematic should show the route of that flow and where it re-enters the main piping. The test data should note which points use the bypass and its magnitude. Notes such as "the bypass valve was opened slightly" or "the valve was wide open" are useful. Secondary flows may not mix thoroughly, or may induce swirl, both of which can affect compressor performance and/or flow meter readings.

406.4 If the compressor is being tested in conjunction with a test of the driver, the requirements for that testing must also be considered. If driver testing requires operation at a single flow point for a long duration, the compressor data may be regarded as a single test point if the readings are stable per code guidelines. As noted in Section 407.1, individual data scans may be treated as data points, and comparisons of the results provide verification of the system stability.

406.5 If the purpose of the test is to verify the guarantee point, then the speed, molecular weight and gas analysis, inlet pressure and temperature must be very close to design. The test flows should span the design point, unless the guaranteed flow can be matched very closely.

406.6 When the compressor manufacturer is involved with the testing, good predictions can be made for test points that do not match the guarantee condition.

407.0 Duration of Test Points

407.1 For this discussion, a test point is defined as "one or more sets of readings of all the instrumentation at a set flow." A "set of readings" may also be called a scan.

407.2 Item 3.12 of PTC-10 states that "the duration of a test point, for a compressor with a constant speed driver, is to be 30 minutes, with no less than five readings of each essential instrument; for compressors with variable speed drivers that require measurement of steam rates, the minimum duration is 60 minutes, with no less than 8 readings of each instrument." Some compressor systems maintain constant conditions for that length of time, but others do not.

407.3 Fewer sets of readings can be used for site testing and shop testing. Treating each set of readings at a given flow setting as a separate test point is preferable to averaging fluctuating data. If possible, at least two test points should be taken at the given flow. If the compressor system is indeed stable or fixed, the data points will plot on top of each other. This avoids the frustration of trying to take steady state points on a system that is drifting.

407.4 Some compressors operate on constantly varying conditions due to steam injection, water spraying, etc. If the system varies at a rapid rate, this should be noted with the test data. "Rapid rate" is difficult to define. Readings of pressure gauges or manometers may show fluctuations that are about a mean level and be steady state readings. Data should be regarded as questionable if the pressures, temperatures, flows or gas compositions are varying noticeably over a 15 - 30 minute period.

408.0 Selecting Instruments and Methods of Measurement

During the test planning period, all methods of measurements should be reviewed for suitability in determining the required results. Instrumentation to be used should be reviewed to see that acceptable levels of accuracy and precision are met. The accuracy, precision and calibration of instruments to be used should be discussed and agreed upon by all parties during the test planning period. Recommended accuracy levels will be listed as they apply to the particular type of test being discussed in each respective section of this procedure. A representative, though not all-inclusive, listing of available instrumentation, can be found in Table 1.

Items that may be measured during a test are: pressure, temperature, gas composition, flow, power, speed and vibrations. No effort will be made to separate the basic variables, such as temperature and pressure, from the derived parameters, such as flow or power.

409.0 Testing Methods

There are many different types of methods available for testing compressors. The choice of method depends firstly on the type of test (performance, mechanical, etc.), and secondarily upon such things as the gas to be tested, the test location or other factors. The methods suitable for a certain type of test will be discussed in that particular section of this procedure.

409.1 Pressure and Temperature Measurement

409.1.1 Pressure gauges should be selected so that measured pressures are near the midpoint of the scale.

409.1.2 When manometers are used, fluid of suitable gravity should be selected such that the minimum fluid level to be read gives the required accuracy. An inclined manometer may be used to increase the accuracy of small pressure differential readings. Total pressure may be obtained either by measuring static pressure and calculating velocity pressure, or by direct measurement with the use of a "total" (pitot) probe.

409.1.3 Static pressure may be measured by means of a pitot static tube, or by drilling a hole in the pipe. If the drilled hole is used, this hole must be smooth and free of burrs.

409.1.4 Pressure measurement must be made at a point of uniform pressure distribution in the flowing stream. If this is not possible, then special pressure profile measuring equipment must be used.

Pressure	Deadweight gauge Bourdon tube gauge Liquid manometer Impact tube Pitot-static tube Pressure transmitter Pressure transducer Barometer
Temperature	Mercury-in-glass thermometer Thermocouple Resistance thermometer
Gas Composition	Gravity balance Psychrometer Dew point apparatus Orsat apparatus Gas chromatograph
Flow	Standardized flow nozzle Orifice plate Venturi tube Displacement meter Elbow meter Rotameter Pitot tube
Power	Wattmeter, ammeter, voltmeter Instrument transformer Torquemeter Dynamometer
Speed	Revolution counter-mechanical Revolution counter-electronic Tachometer Stroboscope
	Kevphasor

Table 1. Instruments organized by property to be measured

409.1.5 When a refrigerant is used as the test fluid, condensation may occur in the high-pressure tubing that connects the probe to the gauge. This is caused by a cool down to ambient temperature of non-flowing gas in this line. This problem can be avoided either by reducing the test pressure levels, by heating all high-pressure tubing, or by mounting the pressure measuring device locally.

409.1.6 The entire pressure measurement system, from probe to readout, must be designed to safely handle the pressure levels to which it will be exposed.

409.1.7 Any calibration to be performed must consider the complete system, from probe to readout device. This is an especially important point when using pressure transmitters or transducers.

409.1.8 As in the case of pressure measurement, total or stagnation temperature can be measured by the use of a "total" temperature probe. The complete temperature measurement assembly must be calibrated, including the probe, reference junction, lead wires and the readout device.

409.1.9 Temperature measurement must be made at a point of uniform temperature distribution in the flowing stream. If this is not possible, then special temperature profile measuring equipment must be employed. Care should be taken that no significant amount of heat can be transferred— by radiation or conduction—to or from the temperature measuring device other than that of the medium being observed.

409.2 Gas Component Measurement

409.2.1 The methods and instrument to be used are a function of the test gas. When air is used as the test medium, a simple measurement of the amount of water vapor present is sufficient. Specific gravity may be measured for other test gases.

409.2.2 If the test gas is a mixture, samples must be taken for analysis by spectrographic, chromatographic or chemical means. This analysis must consist of identification of constituents and a measure of the mole percent of each.

409.2.3 Extreme care must be exercised when obtaining the sample. A clean, well-purged sample bottle must be used. Also, the location for drawing off the sample must be selected with care to ensure that a representative sample is taken for the stage of compression to be analyzed.

409.3 Flow Measurement

409.3.1 The flows to be measured can be liquid, such as lube oil, or gaseous, such as process gas. The same principles hold true for the measurement of either. ASME PTC 19.5 offers a very clear presentation of the principles, installation, equations, and constants necessary to measure flow accurately.

409.3.2 In general, if the device is suitable for the fluid and flow rate to be measured and manufactured to tolerances specified by ASME PTC 19.5, then the most important aspect of flow measurement is correct installation of the flow measuring element.

409.3.3 This is especially true in gaseous flow measurement. The fluid must enter the flow measuring element with a fully developed velocity profile. This condition is best achieved by the use of adequate lengths of straight pipe upstream of the flow element (PTC-19.5). Alternately, flow straighteners can be used.

409.4 Power Measurement

409.4.1 The power input at the compressor shaft may be measured directly by torque meters or reaction mounted drivers, or determined indirectly, from measurements of electrical input to a driving motor, by a heat balance method or by heat energy transferred to a loop cooler.

409.4.2 The most frequent method of power determination is the heat balance, or enthalpy rise method. Accurate measurement of gas composition and conditions at inlet, and discharge of the compression stage are necessary.

409.4.3 Care must be taken that shaft gas seal leakages, bearing friction losses, and heat loss from the casing are all considered These items may not always be of consequence, especially the heat lost from the casing. But, all items should be recognized as being present so that they will be allowed for when they do become consequential. This method of power determination requires very accurate temperature measurement at inlet and discharge of the compression stage.

409.4.4 Before recording any test data, the compressor must be operated at the test flow point for a sufficient amount of time to allow thermal equilibrium to be established throughout the compressor.

409.4.5 Power measurement by means of speed and torque measurement requires special care in the calibration of the torque meter. Calibration must be performed before and after testing. Also, extra care in shaft alignment and coupling balancing are necessary to avoid or limit errors introduced by vibrations.

409.4.6 Electrical power input to a motor can also be used to measure the shaft horsepower of a compressor. Consideration must be given to the efficiency of the motor and the speed increasing gear, if one is involved.

409.4.7 Estimation of these losses at load, even when test data is available for the motor efficiency at no load, make the method somewhat less accurate than the enthalpy rise or torque meter methods.

409.4.8 The heat exchanger method of power measurement is usually not employed and will not be discussed here. If this method is to be used refer to ASME PTC-10.

409.5 Speed Measurement

409.5.1 Where variable speed drivers are used, instruments should be selected to provide a continuous indication of speed fluctuation. Use of two independent instruments, one to provide a check on the other, is recommended. A preferred speed measuring instrument is the electronic counter when actuated by the magnetic pulse generator.

409.5.2 Revolution counters, which use automatic built-in timers based on the frequency of the local A/C power circuit, must be checked against a suitable standard.

409.5.3 Vibrating reed types of tachometers may be used only where their readings can be checked by another instrument that is not sensitive to vibration frequency.

409.5.4 The speed of a compressor driven by synchronous motors may be determined from the number of poles in the motor, and the frequency of the power system.

500.0 TEST PROCEDURES

501.0 Site Test

The purpose of the site performance test is to evaluate the performance of the centrifugal compressor to assure it is operating as guaranteed.

502.0 Test Planning

502.1 Proper planning is essential to the performance of a meaningful site test. An agenda covering the details of the test should be drawn up well in advance, and agreed upon by the manufacturer and the customer.

502.2 The instruments required should be listed, secured, calibrated and placed properly. The data required must be identified.

502.3 Field conditions shall not exceed the allowable departures from the specified operating conditions shown in Table 2.

VARIABLE	UNIT	DEPARTURE %
Inlet Pressure	PSIA	5
Inlet Temperature	°R	8
Specific Gravity of Gas	Ratio	5
Speed	RPM	2
Capacity	CFM	4
Gas Analysis	Mol%	1

Table 2. Allowable Departure from Specified Operating Conditions for Class I Tests (Source: ASME PTC-10)

502.4 Allowable fluctuations of test readings during the duration of a test run are listed in Table 3.

Measurement	Unit	Fluctuation %
Inlet Pressure	PSIA	2
Inlet Temperature	°R	0.5
Discharge Pressure	PSIA	2
Discharge Temperature	°R	0.5
Nozzle or Flow Meter Differences Pressure	PSI	2
Nozzle or Flow Meter Differences Temperature	°R	0.5
Speed	RPM	0.5
Specific Gravity	Ratio	1
Gas Analysis	Mol %	0.01

Table 3. Allowable Fluctuations of Test Readings During a Test Run for All Tests, Class I, II, and III (Source: ASME PTC-10)

503.0 Test Measurements

503.1 The following measurements will be taken: inlet and discharge pressure to each compressor case.

503.2 Two pressure taps or stations, 180 degrees apart, will be used on each inlet and discharge. Location of the pressure taps will be in accordance with Figure 1.

503.3 Inlet and discharge temperature to each compressor case: Two temperature probes or stations, 180 degrees apart will be used on each inlet and discharge. Location of the temperature probes will be approximately 6 inches from the compressor nozzle flanges.

503.4 Inlet flow to each compressor case will be measured by the use of a suitable flow measuring device. The gas should enter the primary element with a fully developed velocity profile, free from swirls or vortices. This condition is best achieved through use of adequate lengths of straight pipe, both proceeding and following the primary element. If it is impossible to use the recommended lengths of straight pipe, straightening vanes or perforated plates should be placed upstream of the measuring device, and the Delta P across the orifice flanges will be measured.

503.5 Gas samples from one of the pressure gauge taps in 503.2 above shall be taken on the suction and discharge of each compressor case at the beginning and end of each test point. The sample will be analyzed by an independent laboratory. The method used to arrive at the properties of the gas must be agreed upon by the manufacturer and customer in advance. Data should be obtained from a flowing well. If that is not possible, a sample should be taken from a deeper well position.

503.6 Speed of compressors, power turbine and gas generator; fuel gas pressure supply to the turbine; gas turbine inlet air temperature; and, gas turbine inlet temperature will all be measured with the appropriate device.

503.7 Gas generator compressor discharge pressure will be measured at the turbine case, and gas turbine inlet filtration Delta P will also be measured with a suitable device.

503.8 Overall, inlet filtration and ducting loss will be measured at the turbine inlet, located within one foot of the inlet protective screen.

503.9 Vibration readings at all monitored points will be recorded and barometric pressure will be recorded before and after each test.

504.0 Test Procedure

The site performance test is executed after all the required instrumentation is selected, installed and calibrated. Operation of the compressor should be stabilized and the variables listed in Section 502.3 must be within the limits stated before accurate readings can be taken.

504.1 Compressors with fixed speed drives are run near their overload condition. By throttling either the suction or discharge valve, the flow is decreased in increments, and measurements are made at each value of flow until a minimum flow operating point is reached.

504.2 Compressors with variable speed drives can vary flow by changing the speed of the compressor. Common procedure, however, is to run at a particular speed and vary flow by throttling.

504.3 After the speed and flow have been set for the first point, the compressor variables should be stabilized and then the various measurements should be recorded.

504.4 After one complete set of data is recorded, a second set must be taken. A minimum of two sets of data should be taken for each point. Readings are taken for flow variations from the maximum flow operating point (close to overload) to the minimum flow operating point (close to surge).

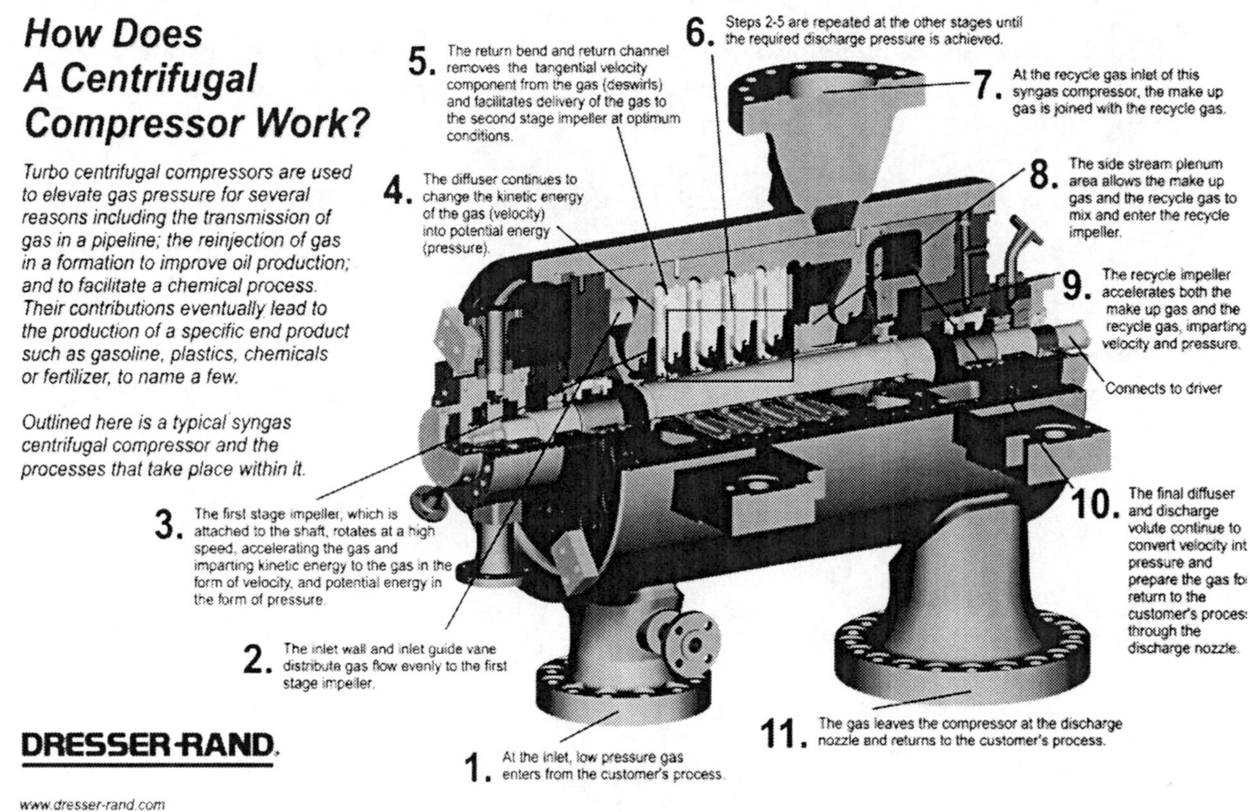

How Does
A Centrifugal
Compressor Work?
Turbo centrifugal compressors are used to elevate gas pressure for several reasons including the transmission of gas in a pipeline; the reinjection of gas in a formation to improve oil production; and to facilitate a chemical process. Their contributions eventually lead to the production of a specific end product such as gasoline, plastics, chemicals or fertilizer, to name a few.
Outlined here is a typical syngas centrifugal compressor and the processes that take place within it.
1. At the inlet, low pressure gas enters from the customer's process.
2. The inlet wall and inlet guide vane distribute gas flow evenly to the first stage impeller.
3. The first stage impeller, which is attached to the shaft, rotates at a high speed, accelerating the gas and imparting kinetic energy to the gas in the form of velocity, and potential energy in the form of pressure.
4. The diffuser continues to change the kinetic energy of the gas (velocity) into potential energy (pressure).
5. The return bend and return channel removes the tangential velocity component from the gas (deswirls) and facilitates delivery of the gas to the second stage impeller at optimum conditions.
6. Steps 2-5 are repeated at the other stages until the required discharge pressure is achieved.
7. At the recycle gas inlet of this syngas compressor, the make up gas is joined with the recycle gas.
8. The side stream plenum area allows the make up gas and the recycle gas to mix and enter the recycle impeller.
9. The recycle impeller accelerates both the make up gas and the recycle gas, imparting velocity and pressure.
Connects to driver
10. The final diffuser and discharge volute continue to convert velocity int pressure and prepare the gas fo return to the customer's proces through the discharge nozzle.
11. The gas leaves the compressor at the discharge nozzle and returns to the customer's process.
DRESSER-RAND.
www.dresser-rand.com

Figure 1. Cross-Section of Centrifugal Compressor and Key Process Points

504.5 When throttling, no liquid must enter the compressor with the throttled gas as this will adversely affect the test results.

504.6 If good data cannot be obtained while the compressor is on process, the compressor can be put in recycle. Flow variation can then be obtained by varying recycle.

504.7 Samples of the gas being compressed, taken at suction and discharge at the beginning and end of each test point, must be analyzed in a laboratory to determine the volumetric percentage of the constituents in the gas. It is very important that the molecular weight does not change during a test.

504.8 Any test point shall last for a minimum of 30 minutes to allow for all temperatures of equipment and piping to stabilize.

600.0 COMPUTATION OF RESULTS

600.1 All acceptable readings shall be averaged and corrected for instrument constants to obtain, for each item observed, a single average value to be used for the computation of results.

600.2 Inlet capacity, polytropic head, efficiency, polytropic gas horsepower, and input shaft horsepower will be calculated.

600.3 Performance evaluation is based on total pressures, so the velocity pressure must be calculated and added to the static pressure to obtain the total pressure. If the velocity pressure is more than 5% of the pressure rise, it is determined by a pitot tube traverse.

600.4 Polytropic exponent is evaluated, and the polytropic efficiency is calculated.

700.00 EVALUATION OF RESULTS

701.00 Guidelines For Field Modifications

701.1 As noted in Section 101.1, a secondary purpose of this procedure is to provide guidelines to help the user to decide when modification should be considered. The general term "modifications," as used herein, includes the following possible procedures:

701.1.1 Mechanical retrofitting having no appreciable effect on hydraulic performance, such as installation of redesigned bearings, seals, couplings, etc.

701.1.2 Peripheral revisions having no appreciable effect on performance, such as modification or addition of lube/seal systems, monitors, instrumentation, or other accessories.

701.1.3 Performance upgrading to improve efficiency and/or handle new conditions. Upgrading could include: minor corrective work entailing no basic change of compressor or driver, such as adjustment or addition of external controls, resetting turbine governor, gear/pinion swap out, etc.

701.1.4 Major revisions, entailing basic modification of compressor or driver, driver replacement, and/or gearbox replacement, etc.

701.2 The specific course of action for major performance upgrading can be established after an evaluation of the following:

- Manufacturer's design data
- Proposed operating conditions (if appreciably different from original design conditions)
- Test results developed in accordance with this procedure
- Operating and maintenance history/problems
- Economic considerations

701.3 Feasibility of performance upgrading would entail correlation of all the items listed above, and estimating probable pay-out time.

701.4 For further delineation of the scope of this section, the following activities, normally classified as maintenance, are not considered to be "modifications:"

- Cleaning of any kind
- Installation of on-stream cleaning provisions
- Replacement of any parts with essentially identical parts
- Repairs intended to restore original capabilities
- Balancing

- Feasibility of mechanical or peripheral improvements having no significant effect on hydraulic performance would probably be determined by the consequences of unscheduled shutdowns and/or shortened machine life, as well as safety considerations.

702.0 Error Analysis With Example

The accuracy of the performance evaluation of a compressor depends upon the validity of physical properties and measured operating conditions used in the calculations. The physical properties of the fluid being compressed are always subject to error. For purposes of this discussion, physical property errors are not treated in this analysis method. The best available source of physical properties should always be used in the calculations. This method will deal with measurement errors in operating conditions of the compressor being tested.

702.1 Given a compressor problem, the engineer must decide an answer for the question: How accurate is compressor power, efficiency, and head computed from possibly imperfect test data? The answer is that a "window of plausible performance" can exist for the most probable range of real performance measurements. The test data may not represent the real performance of a compressor for two reasons:

702.1.1 The finite time lapse in gathering the data may produce an inconsistency in part of the data with respect to the rest because of non-stable compressor operation.

702.1.2 A finite error in the measurement of one or more parameters may arise for various reasons, such as improper test point locations, installation and calibration.

702.2 One straight-forward methods of analysis suggests that the engineer take the test data axis and perform calculations to arrive at a performance point, which would include one efficiency, power requirement and head for the given set of data. Then, one may make various assumptions in individual parameters and repeat the calculations to see what impact the change has on the calculated results. For example, the suction or discharge pressure could be increased or decreased by 0.5% or 1.0%, or the suction or discharge temperature increased or decreased by 1 °F or 3 °F. Anyone performing such tests should vary only one parameter at a time to see its effect on head, horsepower and efficiency.

702.3 Example

Let us assume we have a small centrifugal air compressor that can be retrofitted with different impellers or sets of impellers to achieve different pressure ratios, from 1 up to 4. In all cases, the mass flow is held constant by discharge throttling at 1,000 pounds per hour. Let us assume that the real efficiency of all of the sets of impellers is 70%. One further simplification is that the compressor is pumping dry air (no water saturated in the air). The theoretical performance of the compressor is given in Table 4 below. For illustration, data at 60 and 80% efficiency are also shown

Suction		Pressure Ratio	Discharge		60%E	70%E	80%E
Pressure PSIA	Temperature ° F		Pressure PSIA	Temperature ° F			
14.7	80	1.2	17.64	130.6	123.1		117.6
		1.5	22.05	198.1	180.0		167.7
		2.0	29.40	295.0	260.1		235.3
		3.0	44.10	451.8	388.6		343.2
		4.0	58.80	579.7	490.7		427.3

Table 4. Theoretical performance of dry air compressor

702.3.1 In reviewing the table, the first notable fact is that, for small pressure ratios approaching one, the gas temperature rise becomes very small. To define efficiency accurately within one percentage point at a pressure ratio of 1.2 would require temperature measurement accurate to less than 1 °F. From the table, one can see that, as pressure ratio becomes greater, discharge temperature becomes much greater, allowing more tolerance of less accurate temperature measurement.

702.3.2 If one takes the theoretical performance data for the 70% efficiency impellers and allows the discharge temperature to vary by +1 °F and +3 °F, a graph can be made of the calculated efficiencies showing that, if the error in temperature was zero, then the calculated efficiency would equal the true efficiency of 70%. The curves, denoted as a discharge temperature error of ∀1, ∀3 °F, show how the calculated efficiency deviates from 70% for various pressure ratios. The magnitude of the error in efficiency goes up as the pressure ratio approaches 1. It would not be possible to measure efficiency accurate to less than one percentage point at pressure ratios below 1.5 unless discharge temperature was measured to an accuracy of less than 1 °F error. As pressure ratios become larger, even an ∀ 3 °F error in discharge temperature could be acceptable for efficiency, accurate within one percentage point.

702.3.3 A similar graph can be made for this example illustrating pressure error for 70% efficiency. A ∀ 1% error in discharge pressure would make the calculated efficiency in error greater than one percentage point for pressure ratios less than 2. For pressure ratios of 1-1.5, even ∀0.5% error in discharge pressure would produce efficiency errors greater than one percentage point. Therefore, the accuracy of the pressure or temperature measurements required for a compressor is strongly dependent on the actual pressure ratio.

702.3.4 The temperature rise from inlet to outlet of a compressor at constant pressure ratio becomes smaller as the efficiency increases. A given error in the measured temperature rise becomes a larger percentage of the temperature rise as the efficiency increases. The same logic applies to pressure measurement errors, because the ideal temperature rise is a function of pressure ratio.

702.3.5 The air compressor example, which showed the calculated error in head associated with ∀0.5% and ∀1% error in discharge pressure, leads to the conclusion that pressure measurement more accurate than 0.5% error would be necessary for head error to be less than ∀1% at pressure ratios below 2. A head error of greater than 2% can be expected for compressors with pressure ratios of 1.2 or less, unless pressure measurement accuracy is less than 0.5%.

702.3.6 Errors in pressure measurement will not affect calculated horsepower, because head and efficiency errors cancel in a reciprocal relationship. The ratio of head to efficiency will not change, although both head and efficiency will change with the pressure error introduced.

702.3.7 Errors in temperature do affect the calculated power of the compressor. For the air compressor example, horsepower error increases at lower pressure ratios for different discharge temperatures. At pressure ratios below 2, temperature errors of 3 °F produce horsepower errors of 1.5% or greater. For pressure ratios of 1.5 or less, horsepower errors of greater than ∀ 1% can be expected because of the difficulty in measuring temperature to accuracy of less than 1 °F.

702.3.8 With some care and good instrumentation, freshly calibrated, pressures and temperatures can be measured with acceptable accuracy. Flow rate, on the other hand, is fairly difficult to measure with certainty. The absolute percentage error in flow measurement will be translated directly into a proportionate calculated power error, since power is a linear function of gas flow. For all standard types of meters, corrections must be made to the raw flow reading for actual flowing gas temperature pressure and composition. Under the best of circumstances, it is very difficult to measure the flow of gases accurately.

702.4 Instrument accuracy is beyond the scope of this discussion. However, for the best possible performance test data, the following suggestions can be given as guidelines.

702.4.1 Temperature measurements should be made using National Bureau of Standards special grade thermocouples, or mercury in glass thermometers with certified calibration.

702.4.2 Pressure gauges of precision test quality (∀ 114% F.S. accuracy) should be used. For vacuum readings or positive pressure up to about 20 psig, manometers and electric pressure transducers may be more accurate than dead weight tested gauges.

Printed in the United States of America
ED-02-21-13